3 Pennsylvania PSSA Grade 3 Math Practice Tests

Full-Length Test Prep with Detailed Answer Explanations

Dr. A. Nazari

Published by View Math Education

ViewMath.com

3 Practice Tests to Get You Started!

Hey there, future math whiz! 

This book has **3 full practice tests** to help you warm up for the real thing. Think of it like stretching before a big game — these tests will get your brain ready and show you what to expect!

👍 Three tests is the **perfect start**!

👍 Each one helps you feel **more ready**!

👍 You'll be surprised how much you **already know**!

Sharpen your pencil and let's get warmed up! 

> **"** Three practice tests is a great way to start. Take your time with each one, and you'll feel more confident every step of the way! **"**

📖 How to Use This Book 📖

Your quick-start guide to 3 great practice tests!

☰ What's Inside This Book

- **3 Full-Length Practice Tests** — Each one covers all the Grade 3 math topics you need to know!

- **Answer Key with Explanations** — Find out why each answer is correct, not just what the answer is.

- **Reference Pages** — A math symbols chart and multiplication table you can peek at any time.

- **A Test Tracker** — Write down your scores and watch your confidence grow!

📅 A Simple 3-Test Plan

With just 3 tests, here's a great way to use them:

- **Test 1** **The Warm-Up.** Take this test without a timer. Get comfortable with the question types. Don't worry about your score — just do your best!

- **Test 2** **The Practice Round.** After reviewing Test 1, try this one with a timer (ask a grown-up!). Focus on the topics that were tricky last time.

- **Test 3** **The Real Deal.** Treat this like the actual test: quiet room, timed, no peeking at answers. See how much you've improved!

⬤ Multiple Choice

Pick the **one best answer** from choices A, B, C, or D. Not sure? Cross out the ones you know are wrong, then pick from what's left. That's a smart move!

✏ Short Answer

Write your answer **and** show your work! Even if your final answer isn't right, showing your steps can earn you credit. Use scratch paper if you need more room.

66 After Each Test 99

Flip to the Answer Key and check your work. For every question you got wrong, **read the explanation carefully.** Then write the tricky topics on your Test Tracker page. If you need extra help, grab our **Grade 3 Math Study Guide!**

⭐ ***Fun fact:*** *Three tests is all it takes to see real improvement! Most kids feel way more confident after just a few rounds of practice.* ⭐

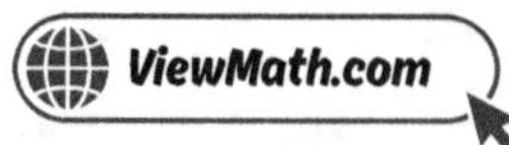

 Tips for Test Day

Easy tricks to help you feel calm and do your best!

☾ The Night Before

- ✓ **Sleep early** — your brain learns while you sleep!
- ✓ **Pack your supplies** — pencils, eraser, scratch paper, all ready to go.
- ✓ **Tell yourself:** *"I've been practicing. I'm going to do great!"*

👍 5 Simple Rules for Every Test

1. **Read the question twice.** The first time to understand it. The second time to catch details.
2. **Show your work.** Write the steps down, even on scratch paper. It helps you think!
3. **Skip the hard ones.** Put a small star next to tricky questions and come back later. Answer the easy ones first!
4. **Never leave a blank.** For multiple choice, your best guess is better than no answer at all.
5. **Check your work.** Finished early? Go back and re-read your answers.

✓ Smart Moves

- Take a deep breath before you begin
- Underline key words in the question
- Use drawings or number lines to help
- Cross out wrong answers first
- Double-check addition and subtraction

✗ Traps to Avoid

- Rushing and not reading carefully
- Picking the first answer that "looks right"
- Forgetting to carry or borrow numbers
- Skipping a question permanently
- Panicking when you see a tough problem

> ❝ Remember, the very first practice test is the hardest — not because the questions are harder, but because everything is new! By Test 3, you'll feel like a pro. Trust me! ❞

🧰 Get Ready to Practice 🧰

Here's everything you need before you start!

Pencils

Sharpened and ready!

Eraser

Everyone makes mistakes!

Scratch Paper

For working things out

A Calm Spot

Somewhere quiet to focus

A Grown-Up

To help set a timer

A Can-Do Attitude

You've totally got this!

✔ Allowed During Tests

- Pencils and erasers
- Blank scratch paper
- The **reference pages** in this book
- A ruler (for measurement questions)

✖ Not Allowed

- Calculators
- Phones, tablets, or computers
- Help from anyone else
- Your study guide (save it for after!)

👥 For Parents & Teachers

- With only 3 tests, **space them at least a week apart**. This gives time to review mistakes before trying the next one.
- Let your child take Test 1 untimed to build familiarity.
- After each test, go through the Answer Key together. Focus on **understanding the "why,"** not just the score.
- If a topic keeps tripping them up, review it in our **Grade 3 Math Study Guide** before the next practice test.
- Celebrate every bit of progress — even getting one more question right is a win!

X^1 Math Reference Sheet X^1

Symbol	Name	What It Means	
$+$	Plus (Add)	Put numbers together.	$3 + 5 = 8$
$-$	Minus (Subtract)	Take away from a number.	$9 - 4 = 5$
$\times$	Times (Multiply)	Add equal groups.	$4 \times 3 = 12$
$\div$	Divide	Split into equal groups.	$12 \div 3 = 4$
$=$	Equals	Both sides are the same.	$2 + 3 = 5$
$>$	Greater Than	The left number is bigger.	$7 > 3$
$<$	Less Than	The left number is smaller.	$2 < 9$
$\frac{1}{2}$	Fraction Bar	Part of a whole.	$\frac{1}{2}$ means 1 out of 2 equal parts

Key Math Words

- **Sum** — the answer when you add
- **Difference** — the answer when you subtract
- **Product** — the answer when you multiply
- **Quotient** — the answer when you divide
- **Factor** — a number you multiply
- **Array** — objects in rows and columns
- **Fraction** — a part of a whole
- **Numerator** — the top number in a fraction
- **Denominator** — the bottom number
- **Equation** — a math sentence with $=$
- **Estimate** — a smart guess, close to the real answer
- **Perimeter** — the distance around a shape
- **Area** — the space inside a shape
- **Rounding** — making a number simpler by going to the nearest ten or hundred

- **Add** (+): in all, total, altogether, combined, sum, both, more

- **Subtract** (−): how many more, how many left, fewer, difference, remain

- **Multiply** (×): each, every, groups of, times, rows of, per

- **Divide** (÷): share equally, split, each group, how many groups, per

Get Online

Find more at
ViewMath.com/PA-Grade3

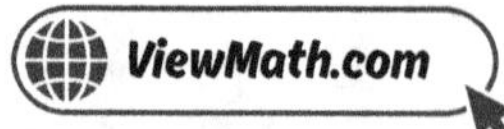

Multiplication Table

×	1	2	3	4	5	6	7	8	9	10	11
1	1	2	3	4	5	6	7	8	9	10	11
2	2	4	6	8	10	12	14	16	18	20	22
3	3	6	9	12	15	18	21	24	27	30	33
4	4	8	12	16	20	24	28	32	36	40	44
5	5	10	15	20	25	30	35	40	45	50	55
6	6	12	18	24	30	36	42	48	54	60	66
7	7	14	21	28	35	42	49	56	63	70	77
8	8	16	24	32	40	48	56	64	72	80	88
9	9	18	27	36	45	54	63	72	81	90	99
10	10	20	30	40	50	60	70	80	90	100	110
11	11	22	33	44	55	66	77	88	99	110	121

💡 How to Use This Table

To find **4 × 7**:

1. Find **4** in the left column (blue).
2. Find **7** in the top row (blue).
3. Follow the row and column until they meet: the answer is **28**!

📈 My Confidence Tracker 📈

My name: _______________________________

☑ Test	📅 Date	⭐ Score	😊 How I Feel
1		/	
2		/	
3		/	

💡 My Quick Reflection

The easiest topic for me was:

____

The trickiest topic for me was:

____

One thing I got better at from Test 1 to Test 3:

____

Next time I want to try:

____

Find more at
ViewMath.com/PA-Grade3

 Table of Contents

Here's what we'll explore together!

 *Let's learn and have fun!*

Practice Test 1

30 Questions

✏️ Before You Start ✏️

- ✔ **Read each question carefully** before choosing your answer.
- ✔ **Show your work** on scratch paper when you need to.
- ✔ **Skip hard questions** and come back to them later.
- ✔ **Check your answers** when you're done.
- ✔ **Take your time** — there's no rush!

⭐ You've Got This! ⭐

Do your best and show what you know!

1. Write 5,600 *in expanded form.*

Your Answer:

2. Which list shows the numbers from greatest to least?

(A) 215, 512, 521

(B) 521, 512, 215

(C) 512, 521, 215

(D) 215, 521, 512

3. What is 268 rounded to the nearest 10?

(A) 260

(B) 270

(C) 300

(D) 200

4. Which statement is true about odd numbers?

(A) Odd numbers always end in 1 or 3

(B) Odd numbers cannot be divided into 2 equal groups

(C) Odd numbers are always less than 100

(D) Odd + Odd = Odd

5. Which number is 10,000 more than 90,000?

(A) 91,000

(B) 99,000

(C) 100,000

(D) 900,000

6. What is the value of the digit 9 in 190,425?

Your Answer:

Find more at
ViewMath.com/PA-Grade3

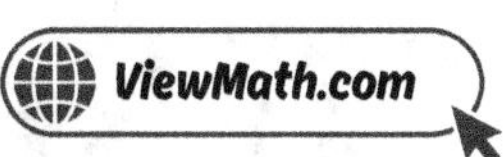

7. A park had 6,238 visitors on Saturday and 4,715 on Sunday. Round each to the nearest 1,000 and estimate the total.

Your Answer:

8. A library has 3,876 fiction books and 4,567 non-fiction books. How many books does the library have in all?

(A) 7,443

(B) 8,333

(C) 8,443

(D) 8,433

9. There are 7 days in a week. How many days are in 9 weeks?

Your Answer:

10. What is 3×200?

(A) 60

(B) 203

(C) 600

(D) 6,000

11. What is the missing number? $? \div 5 = 9$

(A) 4

(B) 14

(C) 45

(D) 54

12. A teacher has 48 books to put on 6 shelves equally. How many books go on each shelf?

(A) 6

(B) 7

(C) 8

(D) 42

Find more at
ViewMath.com/PA-Grade3

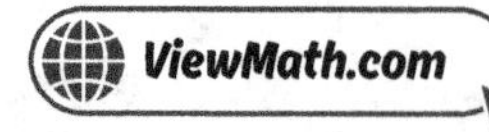

13. A farmer collects 7 eggs each day for 6 days. He uses 10 eggs to bake. How many eggs are left?

Your Answer:

14. Looking at the diagonal of a multiplication table $(1 \times 1, 2 \times 2, 3 \times 3, \ldots)$, what are these numbers called?

(A) Even numbers

(B) Odd numbers

(C) Square numbers

(D) Prime numbers

15. A circle is split into 3 equal parts. None of the parts are shaded. Write the fraction that is shaded.

Your Answer:

16. Would you rather have $\frac{1}{2}$ of a pizza or $\frac{1}{6}$ of the same pizza?

(A) $\frac{1}{6}$ because $6 > 2$

(B) $\frac{1}{2}$ because halves are bigger than sixths

(C) They are the same

(D) It depends on the pizza

17. Count by thirds: $\frac{1}{3}, \frac{2}{3}, ?$. What comes next?

(A) $\frac{3}{6}$

(B) $\frac{3}{3}$

(C) $\frac{4}{3}$

(D) $\frac{3}{9}$

18. Write 1 as a fraction with denominator 8.

Your Answer:

19. Compare: $\frac{3}{4}$ ___ $\frac{3}{8}$

  (A) >

  (B) <

  (C) =

  (D) Cannot compare

20. A library is open from 10:00 A.M. to 3:00 P.M. How long is it open?

  (A) 3 hours

  (B) 4 hours

  (C) 5 hours

  (D) 7 hours

21. An eraser ends at the third quarter-inch mark past 3 inches. How long is the eraser to the nearest quarter inch?

  (A) $3\frac{1}{4}$ inches

  (B) $3\frac{1}{2}$ inches

  (C) $3\frac{3}{4}$ inches

  (D) 4 inches

22. A large milk jug holds 4 liters. You pour out 1 liter for cereal. How much milk is left?

Your Answer:

23. How much is a quarter worth?

  (A) 1 cent

  (B) 5 cents

  (C) 10 cents

  (D) 25 cents

24. You buy a book for $4.50 and pay with a $10 bill. How much change do you get?

  (A) $4.50

  (B) $5.50

  (C) $6.50

  (D) $14.50

Find more at
ViewMath.com/PA-Grade3

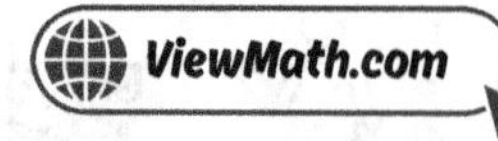

25. Look at this picture graph data. Each symbol stands for 1 pet.

Dogs: ★★★★★★
Cats: ★★★★
Fish: ★★

How many pets are there in all?

(A) 3

(B) 10

(C) 12

(D) 14

26. Using the same rainy days graph (Jan = 6, Feb = 4, Mar = 3, Apr = 7), how many fewer rainy days were in February than in April?

(A) 1

(B) 2

(C) 3

(D) 4

27. A line plot shows pencil lengths in inches. There are 4 X marks above 2 inches and 2 X marks above 3 inches. How many more pencils are 2 inches than 3 inches?

Your Answer:

28. A square has sides that are 7 inches long. What is its area?

(A) 14 sq in

(B) 28 sq in

(C) 42 sq in

(D) 49 sq in

29. Lily says a rectangle that is 5 cm long and 3 cm wide has a perimeter of 15 cm. What mistake did she make?

(A) She subtracted instead of adding.

(B) She found the area instead of the perimeter.

(C) She forgot to add all 4 sides.

(D) She divided instead of multiplying.

Find more at
ViewMath.com/PA-Grade3

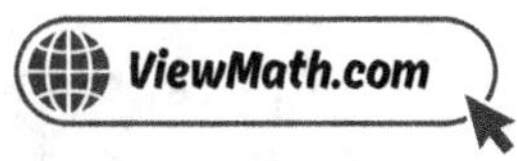

30. If one pizza is cut into 4 equal slices and another identical pizza is cut into 8 equal slices, which slices are bigger?

(A) The $\frac{1}{8}$ slices

(B) The $\frac{1}{4}$ slices

(C) They are the same size.

(D) You cannot tell.

 End of Practice Test 1 

Great job finishing the test!

☑ My Score

I got _____________ out of 30 questions right.

*Check your answers in the **Answer Key** at the back of the book.*

💡 *Review any questions you missed. That's how we learn!*

📊 Check Your Score Online!

Visit **ViewMath Academy** to enter your answers and see which topics you need to review. You can also explore lessons, take quizzes, track your scores, and save your progress!

viewmath.com/score/3.1.PA.01

Or go to *viewmath.com/score* and enter code: *3.1.PA.01*

2

Practice Test 2

 30 Questions

 Before You Start

- ✓ **Read each question carefully** before choosing your answer.
- ✓ **Show your work** on scratch paper when you need to.
- ✓ **Skip hard questions** and come back to them later.
- ✓ **Check your answers** when you're done.
- ✓ **Take your time** — there's no rush!

⭐ You've Got This! ⭐

Do your best and show what you know!

1. Write the expanded form of 9,205.

> Your Answer:

2. Which symbol makes this statement true? 3,482 ______ 3,479

(A) <

(B) =

(C) >

(D) None of these

3. What is 995 rounded to the nearest 10?

(A) 990

(B) 995

(C) 1,000

(D) 900

4. Write the next three odd numbers after 49.

> Your Answer:

5. A city has 85,420 people. What is the value of the digit 5 in this number?

(A) 5

(B) 500

(C) 5,000

(D) 50,000

6. What is the smallest 6-digit number?

> Your Answer:

Find more at
ViewMath.com/PA-Grade3

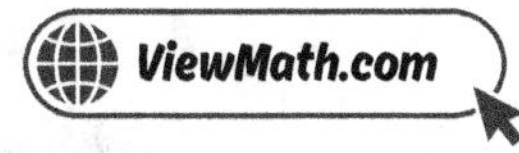

7. Which of these numbers rounds to 7,000 when rounded to the nearest 1,000?

(A) 6,312

(B) 7,499

(C) 7,501

(D) 6,400

8. What is $2,999 + 1,001$?

(A) 3,000

(B) 3,990

(C) 4,000

(D) 4,010

9. What is 8×7?

(A) 48

(B) 54

(C) 56

(D) 63

10. A farmer plants 100 seeds in each row. If there are 7 rows, how many seeds does the farmer plant?

(A) 107

(B) 70

(C) 700

(D) 7,000

11. Solve: $n \div 4 = 7$. What is n?

(A) 3

(B) 11

(C) 24

(D) 28

12. A dog eats 3 treats every day. How many treats does the dog eat in 9 days?

(A) 12

(B) 21

(C) 27

(D) 30

13. Jake has 4 packs of trading cards with 9 cards each. He buys 2 more packs of 9 cards. How many cards does he have now?

(A) 36

(B) 45

(C) 54

(D) 72

14. Look at the ones digits of the multiples of 5: $5, 10, 15, 20, 25, 30$. What pattern do you see?

(A) The ones digit is always 0

(B) The ones digit is always 5

(C) The ones digit alternates between 5 and 0

(D) The ones digit goes $5, 0, 5, 5, 0, 5$

15. Which fraction has a numerator of 2 and a denominator of 6?

(A) $\frac{6}{2}$

(B) $\frac{2}{6}$

(C) $\frac{2}{2}$

(D) $\frac{6}{6}$

16. What is special about ALL unit fractions?

(A) They all have the same denominator

(B) They all equal 1

(C) They all have 1 as the numerator

(D) They are all less than $\frac{1}{2}$

17. $\frac{7}{8}$ is built from how many copies of $\frac{1}{8}$?

(A) 1

(B) 7

(C) 8

(D) 15

18. Write 4 as a fraction with denominator 8.

Your Answer:

19. *Write $<$, $>$, or $=$:* $\frac{1}{4}$ ___ $\frac{1}{8}$

Your Answer:

20. *A TV show starts at 4:15 P.M. and ends at 5:00 P.M. How long is the TV show?*

Your Answer:

21. *Which measurement is the same as $\frac{1}{2}$ inch?*

(A) $\frac{1}{4}$ *inch*   (B) $\frac{2}{4}$ *inch*

(C) $\frac{3}{4}$ *inch*   (D) 1 *inch*

22. *Container A holds 20 liters. Container B holds 13 liters. How much more does Container A hold?*

(A) 7 L   (B) 13 L

(C) 20 L   (D) 33 L

23. *Liam buys an apple for $0.85 and a banana for $0.40. He pays with a $5 bill. How much change does he get?*

(A) \$3.25   (B) \$3.75

(C) \$4.15   (D) \$4.60

24. *Subtract:* $12.00 - \$7.45$

Your Answer:

Find more at
ViewMath.com/PA-Grade3

ViewMath.com

25. Lily says the picture graph below shows that 4 kids were on the swings.

Swings: ★★★★

Slide: ★★★★★★

Key: Each ★ = 2 kids

What is the correct number of kids on the swings?

(A) 2

(B) 4

(C) 6

(D) 8

26. When is a bar graph better to use than a picture graph?

(A) When you have very small numbers

(B) When the numbers are big and drawing many symbols would take too long

(C) When you are measuring lengths

(D) When you only have one category

27. A line plot shows the heights of bean plants in inches. There are 3 X marks above 4, 5 X marks above 5, 2 X marks above 6, and 1 X mark above 7. How many bean plants were measured in all?

Your Answer.

28. A square has sides of 9 inches. What is the area?

Your Answer

29. A triangle has sides of 4 cm, 6 cm, and 8 cm. What is the perimeter?

(A) 14 cm

(B) 18 cm

(C) 20 cm

(D) 24 cm

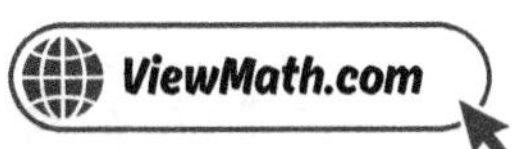

30. *You fold a piece of paper in half, then fold it in half again. How many equal parts do you have?*

Your Answer:

 End of Practice Test 2 

Great job finishing the test!

☑ My Score

I got ____________ out of 30 questions right.

Check your answers in the **Answer Key** at the back of the book.

💡 Review any questions you missed. That's how we learn!

📊 Check Your Score Online!

Visit **ViewMath Academy** to enter your answers and see which topics you need to review. You can also explore lessons, take quizzes, track your scores, and save your progress!

viewmath.com/score/3.1.PA.02

Or go to *viewmath.com/score* and enter code: *3.1.PA.02*

3

Practice Test 3

 30 Questions

- ✓ **Read each question carefully** before choosing your answer.
- ✓ **Show your work** on scratch paper when you need to.
- ✓ **Skip hard questions** and come back to them later.
- ✓ **Check your answers** when you're done.
- ✓ **Take your time** — there's no rush!

 You've Got This! 

Do your best and show what you know!

1. Each place value is how many times bigger than the place to its right?

(A) 2 times   (B) 5 times

(C) 10 times   (D) 100 times

2. How many whole numbers are between 250 and 260, not including 250 and 260?

(A) 8   (B) 9

(C) 10   (D) 11

3. What is 3,421 rounded to the nearest 1,000?

(A) 3,000   (B) 3,400

(C) 3,500   (D) 4,000

4. If you add two odd numbers, the result is always:

(A) Odd   (B) Even

(C) Greater than 10   (D) An odd number less than 20

5. What is the value of the digit 5 in 71,053?

Your Answer:

6. How many place values are used in numbers up to 999,999?

(A) 4   (B) 5

(C) 6   (D) 7

Find more at
ViewMath.com/PA-Grade3

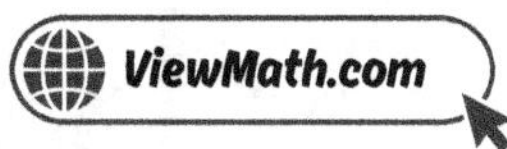

7. What is 5,283 rounded to the nearest 1,000?

(A) 5,000

(B) 5,200

(C) 5,300

(D) 6,000

8. Town A has 4,356 people and Town B has 3,789 people. How many people live in both towns combined?

(A) 7,145

(B) 8,045

(C) 8,145

(D) 8,245

9. The digits of every product in the 9s table add up to what number?

(A) 0

(B) 1

(C) 9

(D) 10

10. Sam says $3 \times 70 = 2,100$. What mistake did Sam make?

(A) He added $3 + 70$ instead of multiplying

(B) He used the wrong basic fact

(C) He added two zeros instead of one

(D) He forgot to add a zero

11. Find the missing number. $8 \times ? = 48$

(A) 5

(B) 6

(C) 7

(D) 8

12. There are 72 crayons shared equally among 8 boxes. How many crayons are in each box?

Your Answer:

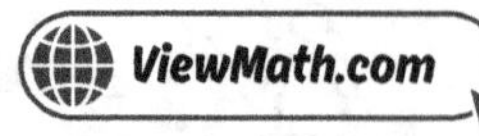

13. A store has 30 red apples and 18 green apples. They put all the apples into bags of 8. How many bags do they fill?

(A) 4

(B) 6

(C) 8

(D) 48

14. In an addition table, what pattern do you see going down the column for $+3$?

(A) Each number increases by 1

(B) Each number increases by 2

(C) Each number increases by 3

(D) Each number stays the same

15. You eat $\frac{2}{6}$ of a cake. How many equal parts does the whole cake have?

Your Answer:

16. Is $\frac{1}{4}$ closer to 0 or to 1 on a number line?

(A) Closer to 0

(B) Closer to 1

(C) Exactly in the middle

(D) At 1

17. To show $\frac{5}{8}$ on a number line, how many hops of $\frac{1}{8}$ do you take from 0?

(A) 3

(B) 5

(C) 8

(D) 13

18. What whole number does $\frac{6}{3}$ equal?

(A) 1

(B) 2

(C) 3

(D) 6

Find more at
ViewMath.com/PA-Grade3

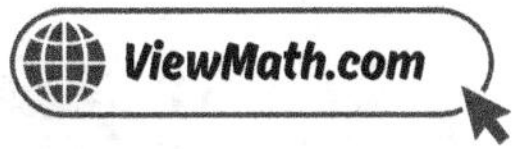

19. When two fractions have the same numerator but different denominators, which fraction is greater?

(A) The one with the bigger denominator

(B) The one with the smaller denominator

(C) They are always equal

(D) You cannot compare them

20. Emma started reading at 3:30 and stopped at 4:10. How long did she read?

(A) 30 minutes

(B) 40 minutes

(C) 50 minutes

(D) 1 hour and 10 minutes

21. A pencil is $7\frac{1}{4}$ inches long. A crayon is $5\frac{3}{4}$ inches long. How much longer is the pencil than the crayon?

Your Answer:

22. What unit do we use to measure liquid volume?

(A) Grams

(B) Kilograms

(C) Liters

(D) Inches

23. How many cents are in $1.00?

(A) 1 cent

(B) 10 cents

(C) 50 cents

(D) 100 cents

24. What is $4.60 + $3.85?

(A) $7.45

(B) $8.45

(C) $8.35

(D) $7.85

Find more at
ViewMath.com/PA-Grade3

25. *A picture graph tracks books read in a month. Each symbol stands for 2 books. Amy has 5 symbols, Ben has 8 symbols, and Chloe has 3 symbols. How many books did they read altogether?*

Your Answer:

26. *A bar graph shows the number of students who play each instrument. Piano = 8, Guitar = 12, Drums = 4, Flute = 6. How many students play Piano or Flute?*

(A) 10
(B) 14
(C) 16
(D) 20

27. *Jake measured 15 sticks. His line plot shows 14 X marks total. What most likely happened?*

(A) He measured one stick twice
(B) He put one X mark in the wrong place
(C) He forgot to plot one measurement
(D) Nothing is wrong

28. *A rectangle is 10 feet long and 3 feet wide. What is its area?*

(A) 13 *sq ft*
(B) 26 *sq ft*
(C) 30 *sq ft*
(D) 33 *sq ft*

29. *Which shape has the greatest perimeter?*

(A) *A square with sides of 5 cm*
(B) *A rectangle that is 8 cm × 3 cm*
(C) *A rectangle that is 6 cm × 4 cm*
(D) *A triangle with sides 7 cm, 7 cm, and 7 cm*

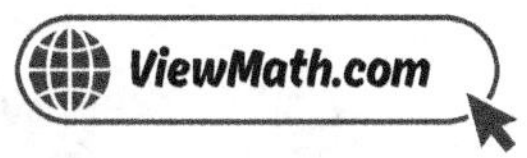

30. A square is partitioned into 4 equal parts. All 4 parts are shaded. What fraction is shaded?

(A) $\frac{1}{4}$

(B) $\frac{3}{4}$

(C) $\frac{4}{4}$

(D) $\frac{4}{1}$

Find more at
ViewMath.com/PA-Grade3

 End of Practice Test 3 

Great job finishing the test!

 My Score

I got _____________ out of 30 questions right.

*Check your answers in the **Answer Key** at the back of the book.*

Review any questions you missed. That's how we learn!

Check Your Score Online!

Visit **ViewMath Academy** to enter your answers and see which topics you need to review. You can also explore lessons, take quizzes, track your scores, and save your progress!

viewmath.com/score/3.1.PA.03

Or go to *viewmath.com/score* and enter code: 3.1.PA.03

Answer Key & Explanations

 **Check Your Answers!** 

First try each test on your own, then look here to check.

Read the explanations to learn from any mistakes ⭐

Practice Test 1 — Answer Key

     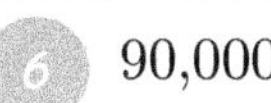 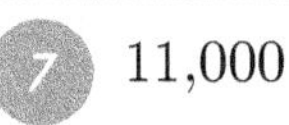 

| 1 | $5{,}000 + 600$ | 2 | B | 3 | B | 4 | B | 5 | C | 6 | 90,000 | 7 | 11,000 | 8 | C |

         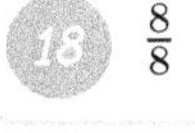

| 9 | 63 | 10 | C | 11 | C | 12 | C | 13 | 32 | 14 | C | 15 | $\frac{0}{3}$ | 16 | B | 17 | B | 18 | $\frac{8}{8}$ |

| 19 | A | 20 | C | 21 | C | 22 | 3 L | 23 | D | 24 | B | 25 | C | 26 | C | 27 | 2 | 28 | D |

| 29 | B | 30 | B |

 **Time to Learn!** 💡

Go through the explanations below, **especially for the questions you missed**.

Understanding why each answer is correct makes you a stronger math thinker!

👍 **Tip:** Circle any questions you got wrong, then read their explanation carefully.

Practice Test 1 — Detailed Explanations

1. $5{,}600 = 5{,}000 + 600 + 0 + 0$. *The tens and ones are both 0.*

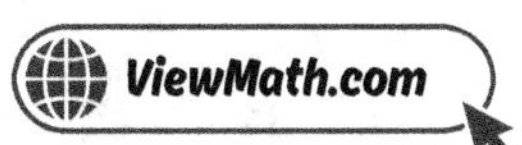

(2) $521 > 512 > 215$. *Comparing hundreds: $5 > 2$, and between 521 and 512, tens: $2 > 1$.*

(3) *The ones digit is 8. Since $8 \geq 5$, round up. $268 \approx 270$.*

(4) *Odd numbers always have 1 left over when divided by 2. They end in $1, 3, 5, 7,$ or 9 (not just 1 and 3).*

(5) $90{,}000 + 10{,}000 = 100{,}000.$

(6) *In $190{,}425$, the digit 9 is in the ten-thousands place, so its value is $90{,}000$.*

(7) $6{,}238 \approx 6{,}000$ *(hundreds digit $2 < 5$). $4{,}715 \approx 5{,}000$ (hundreds digit $7 \geq 5$). Estimated total: $6{,}000 + 5{,}000 = 11{,}000$.*

(8) $3{,}876 + 4{,}567 = 8{,}443$. *Ones: $6 + 7 = 13$, carry 1. Tens: $7 + 6 + 1 = 14$, carry 1. Hundreds: $8 + 5 + 1 = 14$, carry 1. Thousands: $3 + 4 + 1 = 8$.*

(9) $9 \times 7 = 63$ *days.*

(10) *Basic fact: $3 \times 2 = 6$. The 200 has two zeros, so add two zeros: 600.*

(11) *To find the dividend, multiply: $9 \times 5 = 45$. The missing number is 45.*

(12) *48 books shared equally on 6 shelves: $48 \div 6 = 8$ books per shelf.*

(13) *Step 1: $7 \times 6 = 42$ eggs. Step 2: $42 - 10 = 32$ eggs left.*

(14) *Numbers like $1, 4, 9, 16, 25, 36, \ldots$ where a number is multiplied by itself are called square numbers.*

Find more at
ViewMath.com/PA-Grade3

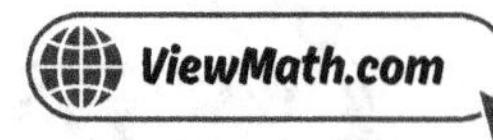

15 0 *parts shaded out of 3 total parts* $= \frac{0}{3} = 0$.

16 $\frac{1}{2}$ *is a bigger piece than* $\frac{1}{6}$ *because splitting into 2 parts gives bigger pieces than splitting into 6 parts.*

17 $\frac{2}{3} + \frac{1}{3} = \frac{3}{3}$ *(which equals 1 whole).*

18 $1 = \frac{8}{8}$ *because 8 eighths make 1 whole.*

19 *Same numerator (3 pieces). Fourths are bigger than eighths, so* $\frac{3}{4} > \frac{3}{8}$.

20 *From 10:00 A.M. to 3:00 P.M.: count* $10 \to 11 \to 12 \to 1 \to 2 \to 3$, *which is 5 hours.*

21 *The third quarter-inch mark past 3 is* $3\frac{3}{4}$ *inches.*

22 $4 - 1 = 3$ *L.*

23 *A quarter is worth 25 cents.*

24 $\$10.00 - \$4.50 = \$5.50$.

25 *Dogs: 6, Cats: 4, Fish: 2. Total:* $6 + 4 + 2 = 12$ *pets.*

26 $7 - 4 = 3$ *fewer rainy days in February than April.*

27 $4 - 2 = 2$ *more pencils are 2 inches long.*

28 *A square is a rectangle with all sides equal. Area* $= 7 \times 7 = 49$ *sq in.*

Find more at
ViewMath.com/PA-Grade3

29　Lily multiplied $5 \times 3 = 15$, which gives the area, not the perimeter. The correct perimeter is $5+3+5+3 = 16$ cm.

30　The $\frac{1}{4}$ slices are bigger because the pizza is cut into fewer pieces. More equal parts means smaller pieces: $\frac{1}{4} > \frac{1}{8}$.

✅ Practice Test 2 — Answer Key

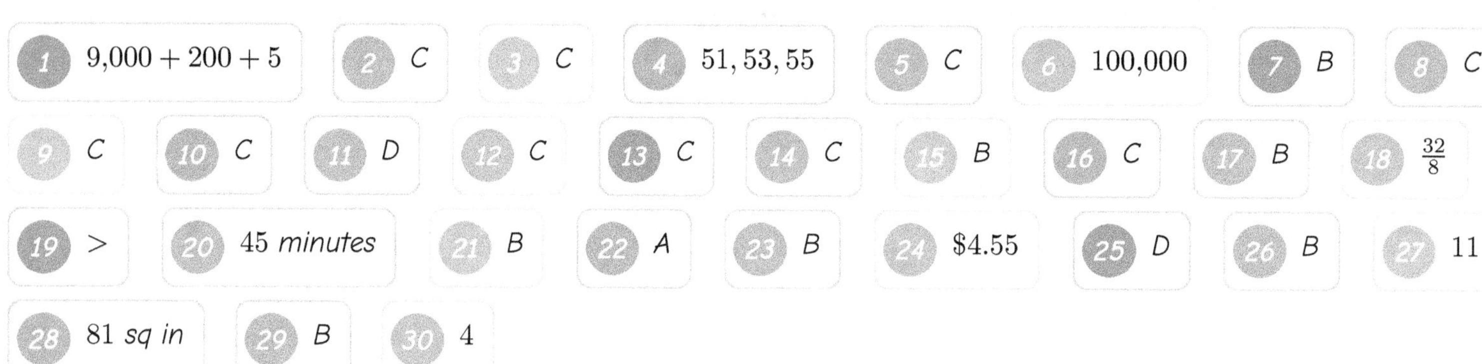

1　$9,000 + 200 + 5$　　2　C　　3　C　　4　$51, 53, 55$　　5　C　　6　$100,000$　　7　B　　8　C

9　C　　10　C　　11　D　　12　C　　13　C　　14　C　　15　B　　16　C　　17　B　　18　$\frac{32}{8}$

19　$>$　　20　45 minutes　　21　B　　22　A　　23　B　　24　$\$4.55$　　25　D　　26　B　　27　11

28　81 sq in　　29　B　　30　4

💡 Time to Learn! 💡

Go through the explanations below, **especially for the questions you missed**.

Understanding why each answer is correct makes you a stronger math thinker!

👍 **Tip:** Circle any questions you got wrong, then read their explanation carefully.

📖 Practice Test 2 — Detailed Explanations

1　$9,205 = 9,000 + 200 + 0 + 5$. The tens digit is 0, so there is no tens term.

2　Both have 3 thousands and 4 hundreds. Comparing tens: $8 > 7$, so $3,482 > 3,479$.

Find more at
ViewMath.com/PA-Grade3

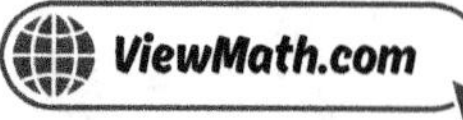

3 The ones digit is 5. Since $5 \geq 5$, round up. $995 \approx 1,000$.

4 Starting from 49, add 2 each time: $49 + 2 = 51$, $51 + 2 = 53$, $53 + 2 = 55$.

5 In 85,420, the digit 5 is in the thousands place, so its value is 5,000.

6 The smallest 6-digit number is 100,000. Any number below it (like 99,999) has only 5 digits.

7 7,499: the hundreds digit is $4 < 5$, so it rounds down to 7,000. The other choices round to 6,000 or 8,000.

8 Ones: $9 + 1 = 10$, carry 1. Tens: $9 + 0 + 1 = 10$, carry 1. Hundreds: $9 + 0 + 1 = 10$, carry 1. Thousands: $2 + 1 + 1 = 4$. The sum is 4,000.

9 $8 \times 7 = 56$.

10 $7 \times 100 = 700$. To multiply by 100, add two zeros.

11 Multiply to find the dividend: $7 \times 4 = 28$. So $n = 28$.

12 $3 \times 9 = 27$ treats.

13 Step 1: Total packs $= 4 + 2 = 6$. Step 2: $6 \times 9 = 54$ cards.

14 Ones digits: $5, 0, 5, 0, 5, 0$. They alternate between 5 and 0.

15 Numerator is on top and denominator is on the bottom. So it is $\frac{2}{6}$.

16 A unit fraction always has 1 as the numerator. Examples: $\frac{1}{2}$, $\frac{1}{3}$, $\frac{1}{4}$, $\frac{1}{8}$.

Find more at
ViewMath.com/PA-Grade3

17 $\frac{7}{8} = 7$ copies of $\frac{1}{8}$.

18 $4 \times 8 = 32$ eighths. So $4 = \frac{32}{8}$.

19   Same numerator. Fourths are bigger pieces than eighths. $\frac{1}{4} > \frac{1}{8}$.

20   From 4:15 to 5:00: count 45 minutes forward. Or $60 - 15 = 45$ minutes.

21 $\frac{2}{4} = \frac{1}{2}$, so $\frac{2}{4}$ inch is the same as $\frac{1}{2}$ inch.

22 $20 - 13 = 7$ L.

23   Total: $\$0.85 + \$0.40 = \$1.25$. Change: $\$5.00 - \$1.25 = \$3.75$.

24 $\$12.00 - \$7.45 = \$4.55$. Borrow to subtract 45 cents from 0 cents.

25   Lily forgot to multiply by the key. Each star $= 2$ kids, and there are 4 stars: $4 \times 2 = 8$ kids.

26   Bar graphs are better for bigger numbers because you don't need to draw many individual symbols — just one bar to the right height.

27 $3 + 5 + 2 + 1 = 11$ bean plants.

28   Area $= 9 \times 9 = 81$ sq in.

29   Add all side lengths: $4 + 6 + 8 = 18$ cm.

30   Folding in half gives 2 parts. Folding in half again doubles it to 4 equal parts. Each part is $\frac{1}{4}$ of the whole.

Find more at
ViewMath.com/PA-Grade3

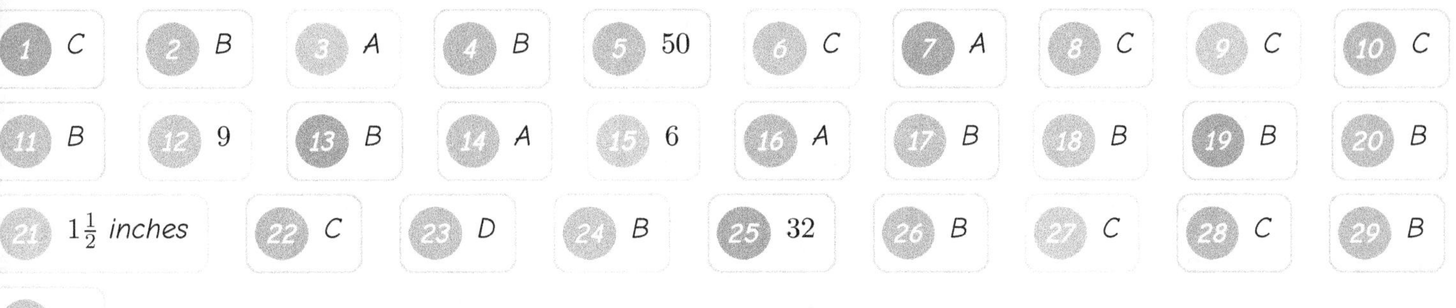

✅ Practice Test 3 — Answer Key

1 C	2 B	3 A	4 B	5 50	6 C	7 A	8 C	9 C	10 C
11 B	12 9	13 B	14 A	15 6	16 A	17 B	18 B	19 B	20 B
21 $1\frac{1}{2}$ inches	22 C	23 D	24 B	25 32	26 B	27 C	28 C	29 B	
30 C									

💡 Time to Learn! 💡

*Go through the explanations below, **especially for the questions you missed**.*

Understanding why each answer is correct makes you a stronger math thinker!

👍 ***Tip:*** *Circle any questions you got wrong, then read their explanation carefully.*

📖 Practice Test 3 — Detailed Explanations

1. Each place is 10 times bigger than the place to its right: ones → tens → hundreds → thousands.

2. The numbers are $251, 252, 253, 254, 255, 256, 257, 258, 259$. That is 9 numbers.

3. Look at the hundreds digit: 4. Since $4 < 5$, round down. $3{,}421 \approx 3{,}000$.

4. Odd + Odd = Even. For example, $3 + 5 = 8$ (even) and $7 + 9 = 16$ (even).

5. In $71{,}053$, the digit 5 is in the tens place, so its value is 50.

Find more at
ViewMath.com/PA-Grade3

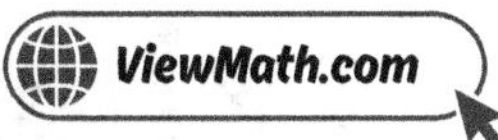

6   Numbers up to 999,999 use 6 places: hundred-thousands, ten-thousands, thousands, hundreds, tens, and ones.

7   The hundreds digit is 2. Since $2 < 5$, round down. $5{,}283 \approx 5{,}000$.

8 $4{,}356 + 3{,}789 = 8{,}145$. Ones: $6 + 9 = 15$, carry 1. Tens: $5 + 8 + 1 = 14$, carry 1. Hundreds: $3 + 7 + 1 = 11$, carry 1. Thousands: $4 + 3 + 1 = 8$.

9   The digits of multiples of 9 always add up to 9. For example, $9 \times 4 = 36$ and $3 + 6 = 9$.

10   The correct answer is $3 \times 70 = 210$. Basic fact: $3 \times 7 = 21$, then add one zero: 210. Sam added two zeros instead of one, getting $2{,}100$.

11 $8 \times 6 = 48$. The missing number is 6.

12 $72 \div 8 = 9$ crayons in each box.

13   Step 1: $30 + 18 = 48$ apples. Step 2: $48 \div 8 = 6$ bags.

14   Going down a column in an addition table, as the row number increases by 1, each sum increases by 1.

15   The denominator tells the total equal parts. The cake is cut into 6 equal parts.

16 $\frac{1}{4}$ is only 1 hop from 0 out of 4 hops total, so it is much closer to 0.

17 $\frac{5}{8}$ means 5 copies of $\frac{1}{8}$, so take 5 hops from 0.

18 $\frac{3}{3} = 1$, so $\frac{6}{3} = 2$.

Find more at
ViewMath.com/PA-Grade3

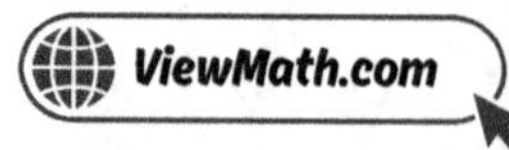

19 Same numerator means the same number of pieces. A smaller denominator means bigger pieces, so the fraction with the smaller denominator is greater.

20 From 3:30 to 4:00 is 30 minutes. From 4:00 to 4:10 is 10 minutes. Total: $30 + 10 = 40$ minutes.

21 $7\frac{1}{4} - 5\frac{3}{4} = 6\frac{5}{4} - 5\frac{3}{4} = 1\frac{2}{4} = 1\frac{1}{2}$ inches.

22 We measure liquid volume in liters (L).

23 $\$1.00 = 100$ cents.

24 Cents: $60 + 85 = 145$ cents $= 1$ dollar and 45 cents. Dollars: $4 + 3 + 1 = 8$. Answer: $\$8.45$.

25 Amy: $5 \times 2 = 10$. Ben: $8 \times 2 = 16$. Chloe: $3 \times 2 = 6$. Total: $10 + 16 + 6 = 32$ books.

26 Piano (8) + Flute $(6) = 14$ students.

27 He measured 15 sticks but only plotted 14 X marks, so he forgot to plot one measurement.

28 Area $= 10 \times 3 = 30$ sq ft.

29 Square: $4 \times 5 = 20$ cm. Rectangle 8×3: $(2 \times 8) + (2 \times 3) = 22$ cm. Rectangle 6×4: $(2 \times 6) + (2 \times 4) = 20$ cm. Triangle: $7 + 7 + 7 = 21$ cm. The 8×3 rectangle has the greatest perimeter of 22 cm.

30 All 4 out of 4 parts are shaded, so $\frac{4}{4}$ is shaded. $\frac{4}{4} = 1$ whole.

Great job checking your work!

Find more at
ViewMath.com/PA-Grade3

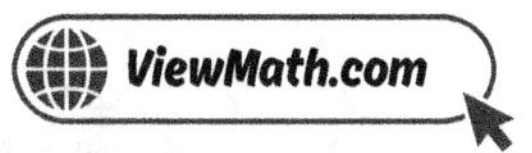

Keep practicing and you'll be a math star!

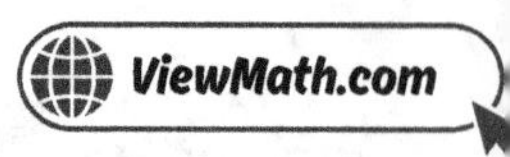